chickens

how to keep chickens
at home

C000240966

Written by
Victoria Roberts BVSc MRCVS

chickens

how to keep chickens
at home

Written by
Victoria Roberts BVSc MRCVS

Magnet & Steel Ltd

www.magnetsteel.com

Printed by Printworks Global Ltd. London/Hong Kong

ISBN: 978-1-907337-21-5
ISBN: 1-907337-21-0

Contents

Introduction

Introduction

Chicken provides 20 per cent of the world's animal protein at a reasonable price, so the human race owes a huge debt to the humble domestic fowl.

But I want to know where my eggs have come from. Wouldn't you like the thrill of producing and eating your own free-range fresh eggs – the finest quality protein which comes in its own packaging?
The taste and texture is something everyone should be able to experience.

Keeping chickens is easy, if you follow a few simple guidelines. But, as with all livestock, you have a duty of care for their health. The responsibility for their welfare is yours, and in the UK this is a legal as well as a moral duty.

Welfare needs

The five welfare needs

- **Environment:** a suitable place to live.

- **Diet:** the right food in the right quantity.

- **Behaviour:** being able to behave normally.

- **Company:** chickens need to live together.

- **Health:** protecting your chickens from pain, suffering, injury and disease.

Before you decide to buy your chickens, you must first consider whether you can meet their five welfare needs.

Environment

Chicken housing is used by the birds for roosting, laying and shelter. Movable pens are good, as the birds get fresh ground regularly. Ventilation is important but needs to be high up, as chickens will not thrive in draughts.

Pictured: Black Rock hybrids.

Diet

Balanced commercial chicken feed is easy to obtain and should be stored dry. Wild birds should be discouraged from accessing chicken feeders.

Behaviour

The pecking order is very important to maintain the stability of a flock and only changes if hens are added or removed. Chickens love to scratch with their feet to find insects – even tiny chicks will do this. The flock keeps watch for potential aerial predators and has a special call to alert friends.

Company

You can keep just two hens together, but chickens are are happy with more company as long as the flock members do not change. They will boss newcomers, for example (see page 90).

Health

It is your responsibility to maintain the health of your birds by fulfilling the conditions listed above, plus keeping stress low and worming regularly.

Chickens take up little time and will benefit a garden with manure as well as being a constant source of fun and entertainment for children and adults alike.

Opposite: A cross-bred bantam hen.

I have been keeping them for nearly 40 years, but still find myself fascinated by the behaviour of these amazing birds.The ancestor of chickens is a small pheasant from Asia, the Red Jungle Fowl (Gallus gallus, hence the general term galliform family) and for the past four thousand years chickens have provided us with eggs, fresh meat and feathers, plus some truly horrible traditional medicines.

There is a huge range of breeds of chickens, plus the hybrids which are commercially selected for high egg production. You may have heard the term poultry – this covers chickens (large fowl and bantams), turkeys, ducks and geese, all of which have been domesticated and are kept for eggs and/or meat.

One small warning – this poultry-keeping hobby is fun but also addictive, and you will not only want more breeds but will wish to hatch them for the adorable chicks. All of this, plus health, welfare and husbandry you can learn in this book.

Opposite: Birchen Pekin bantam hen.

What is a chicken?

Pictured: A pullet
(young female chicken).

What is a chicken?

Birds have several striking differences to mammals in the way that their bodies work. No teeth, the ability to fly, a one-way breathing system and feathers are just some of them.

The head of a chicken has a fleshy part of skin on top called the comb (larger in the male and normally bright red in good health), plus similar skin wattles under the beak. The beak is made of hard keratin that continually grows. There are various types of comb, the most common being the single comb.

The eye is large and blinking is done by the nictitating membrane, a transparent film that travels across from inner to outer eye. When the bird sleeps, the bottom lid closes upwards. The ear canal is short, covered with small feathers. It has a dangling earlobe beneath and is located just behind the eye.

Depending on breed, the earlobe can be red, white (denotes a white-shelled egg) or blue. The nostrils are on the top of the beak and the sinuses are behind these. Most breeds have a red-coloured face and comb, with feathers over the eye ridge but not on the facial skin. Some breeds have dark skin.

All birds have some special flight adaptations, such as hollow bones for lightness, fused vertebrae in the thoracic area (protection of the lungs) and fused lumbar area for protection of the kidneys and egg production, plus ease of walking. A single vertebra between the fused areas acts as a hinge so the bird can bend. The neck is long (16 vertebrae as opposed to 7 in mammals) and very flexible, so that the chicken can preen. This means regularly grooming its feathers, which is so important for keeping warm. The good structure of a feather, with the barbs locked (think velcro), maintains the insulating properties and waterproofing.

Bird respiration works on a circular system (think bagpipes - birds do not need to stop for breath when singing) with the air going one way round. The lungs do not expand, but the air is drawn through them and the oxygen removed more efficiently than in mammals, due to the energy-expensive activity of flight. The pressure is maintained by several pairs of airsacs, which are the clarity and strength of clingfilm.

The lungs, airsacs and hollow bones are all connected, so any respiratory disease can be difficult to control.

Most chickens have feather-free scaled legs, which should be smooth. The scales also moult at the same time as the feathers. Some breeds have feathers growing on the outside of the leg and outer toe, generally having three toes pointing forward and one back, but the Dorking and four other breeds have an extra toe pointing up the back of the leg.

Basic anatomy

'As rare as hens' teeth' is a common saying – true, as food is ground up in the very muscular gizzard instead, with the aid of small stones which the hen has previously eaten. Food and plant material is delicately picked up with the beak (made of keratin, rather like our own nails), texture being more important than taste, while hens use their feet to scratch the soil and expose insects. Colour vision and shadows help the hen to find food. The crop is used to store food which is then passed to the proventriculus (acid-producing stomach) and then to the gizzard where some grinding up is done, passing back and forth several times between the gizzard and proventriculus and then when fine enough, through into the small intestine.

Nutrients are absorbed from the small and large intestine and then the liver filters nutrients and toxins from the food, carried in the blood.

The caeca (paired, blind-ended portions of the gut) is where a certain amount of fermentation of plant material takes place and the droppings from this area are voided at a ratio of about one in ten with the normal droppings. Caecal droppings are slightly frothy and paler in colour than normal droppings, which are brown with a white tip. The white tip is the urates (bird form of urine) which in birds is normally solid.

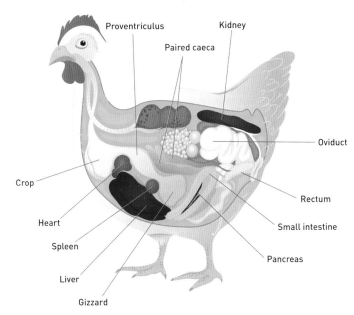

Proventriculus

Kidney

Paired caeca

Oviduct

Crop

Rectum

Heart

Small intestine

Spleen

Pancreas

Liver

Gizzard

See the diagram for the location of internal organs (see page 117). A common mistake among novices is to feel the hard gizzard and think the hen is egg-bound (an egg stuck in the oviduct).

Signs of a healthy chicken

- Dry nostrils

- Red comb (some breeds have naturally dark ones)

- Bright eyes (colour varies with breed)

- Shiny feathers (all present)

- Good weight and musculature for age

- Clean vent feathers with no smell

- Smooth shanks

- Straight toes

- Bird alert and active

Pictured: A healthy head.

Which type will suit you?

You need to decide your priorities before you acquire your chickens. Do you want lots of eggs each week, meat, beauty, or to conserve rare breeds? Do you want pets or the vegetable garden weeded? No single type or breed will fulfill all of these requirements and it is so important that you like the look of the hens you buy – they will become part of the family.

Hybrids are commercial crossbreeds and were originally developed and selected in the 1950s for the battery cage egg industry, to increase egg production. They are based on just a few of the more productive pure breeds and tend to be brown in colour, and uniform in shape and size. Examples are available as Warrens, Isabrowns, and Hy-lines.

These are the cheapest, as they are reared in large numbers. Beware of very cheap hens. These may be ex-battery hens at the end of their laying life and rescuing them, although admirable, will lead to heartache as they sicken and die within a few months.

Opposite: Hybrids: Warren and two Speckledys.

The best production you can expect is 250-300 eggs per year for almost two years.

The original outdoor hybrid, developed for free-range systems with hardiness, and good feathering, is the Black Rock, which has a life expectancy of around four years. There are now several other species available, including Bovans Nera, Calder Ranger, Speckledy, Columbian Blacktail, White Star, and BlueBelle. They are more productive than pure breeds, hardier than the commercial hybrids and provide different colours of birds or eggs. They are a little more expensive than the brown hybrids. Best production is 250-275 eggs per year for three-four years.

Pure breeds are the traditional breeds of poultry, developed in various countries for a variety of purposes, mostly since Victorian times. They are also used for exhibition and have Standards for shape and colour. On the whole they only lay eggs during the longer days. They can produce meat, and some of the rarer breeds are particularly beautiful. They live and lay eggs for four to seven years. The light breeds will lay the most, but can be flighty and nervous. The heavy breeds lay less and eat more, but will produce meat as well. Pure breeds are the most expensive. Expect 100 to 250 eggs per year, depending on breed and age.

Bantams lay small eggs but eat less, and are great for children to keep and look after. Pure breeds are exhibited at shows around the country.

Opposite: A winning combination.

Choosing the right pure breed

The most popular breeds worldwide are the Light Sussex (UK origin, white with black neck, wings and tail) and Rhode Island Red (USA origin, dark red). These are known as dual purpose – that is, they are kept both for eggs and meat.

For laying purposes, any of the Leghorn (Italian origin) colours are popular. For those who think that brown eggs are best, the Marans (French origin) and the Welsummer (Dutch origin) lay the darkest brown eggs. Light breeds are the best layers, heavy breeds tend to be more docile.

Pictured: Light Sussex.

Heavy breeds

These include Marans (two-tone grey banding across the feathers known as Cuckoo) and the Welsummer (typical orange and black farmyard storybook cockerel colour) whose eggs are slightly redder – more of a terracotta colour. A light brown egg is laid by the Barnevelder, the plumage being mahogany with double black lacing on each feather.

Of the British heavy breeds, one of the most popular is the Buff Orpington (not a good layer), which were owned by Queen Elizabeth the Queen Mother. The Sussex is a good egg layer, the most popular colour being Light (white with black points).

The Rhode Island Red, Australorp, Plymouth Rock, and Wyandotte, all lay tinted eggs, while the Croad Langshan lays a plum-coloured egg. The heavier breeds include the Dorking (with five toes) and the Indian Game, which is particularly broad and heavy. The heavy breeds with feathered legs, such as the Cochin, Brahma and Faverolles, lay fewer eggs.

Opposite:
Welsummer hens.

Light breeds

Nearly all of these lay white or light-coloured eggs.
The White Leghorn still out-produces most breeds.
Other Mediterranean breeds are the Ancona (white
spots on black), Minorca and Andalusian (blue
laced). British breeds include Derbyshire Redcap,
Old English Pheasant Fowl, Hamburg (pencilled
or spangled), Scots Dumpy (short legs) and Scots
Grey, all of which have good utility attributes. The
Crested Breeds include the Poland, the Araucana
(blue/green eggs) and that most fluffy of birds, the
Silkie. The Frizzle looks strange with its backward-
curling feathers, but is a decent layer. Old English
Game and the Modern Game are particularly hardy
and colourful.

Large fowl bantams and true bantams

There are miniatures of certain large fowl, which
should be one quarter the size of the large, usually
referred to as bantams. True bantams do not have
a large fowl counterpart and are primarily for
ornamental purposes, but are excellent for
young children and those without much space
to keep poultry.

laying capabilities

A guide to pure breed chicken expected laying capabilities (large fowl).

Breed	Egg colour	Numbers p.a.	Maturing	Type
Ancona	white	200	quick	light
Andalusian	white	200	quick	light
Araucana	blue/green	150	quick	light
Australorp	tinted	180	medium	heavy*
Barnevelder	light brown	180	medium	heavy
Brahma	tinted	150	slow	heavy*
Campine	white	200	quick	light
Cochin	tinted	100	slow	heavy*
Croad Langshan	brownish	180	medium	heavy*
Dorking	white	190	medium	heavy*
Faverolles	tinted	180	medium	heavy*
Fayoumi	tinted	250	quick	light
Friesian	white	230	quick	light
Frizzle	tinted	175	medium	heavy
Hamburg	white	200	quick	light
Indian Game	tinted	100	medium	heavy
Leghorn	white	240	quick	light
Marans	dark brown	200	medium	heavy*
Minorca	white	200	medium	light
Old English Game	tinted	200	quick	heavy*
OE Pheasant Fowl	white	200	quick	light
Orpington	tinted	180	medium	heavy*
Plymouth Rock	tinted	200	medium	heavy
Poland	white	200	quick	light

Derbyshire Redcap	tinted	200	quick	light
Rhode Island Red	tinted/brown	260	medium	heavy
Scots Dumpy	tinted	180	medium	heavy*
Scots Grey	tinted	200	quick	light
Sicilian Buttercup	white	180	quick	light
Silkie	tinted	150	quick	light*
Spanish (white faced)	white	200	quick	light
Sumatra	tinted	200	quick	light*
Sussex	tinted	260	medium	heavy
Welsummer	dark red-brown	200	medium	heavy
Wyandotte	tinted	200	medium	heavy

most likely to go broody. Some colour varieties of breeds lay better than others and different exhibition and utility strains exist, exhibition strains not being as productive.

Pictured: Red Dorking hen.

Study the positive signs of health (see page 24) and follow the biosecurity guidelines (page 122) so that you know what to look for when buying birds. Reject any which do not come up to scratch. Don't buy them because you feel sorry for them - they will be nothing but trouble.

Point-of-lay (POL) is about 18 weeks of age with the fast-maturing hybrids. POL with pure breeds could be 26 weeks or more. It is better to obtain the hens before they begin laying, so that the laying cycle is not upset by a change of environment. If you transport a bird which is in lay, it is very likely to stop laying for a few weeks.

Opposite: POL hens.

Pullets (hens which have not yet laid an egg) may be obtained at an earlier age but will need a grower ration until 18 weeks of age in order for them to grow properly.

Good sources

From advertisements in specialist smallholding magazines.

From private breeders who exhibit at poultry shows.

At small sales, where you can talk to the breeder.

Pet stores, where you can expect advice and health guarantees.

Sources to avoid

Large sales. Prices can escalate and there may not be any history with the birds to let you know how old they are, their health status or how they have been reared.

- Advertisements in local newspapers.

- Car boot sales.

- The internet.

Pictured: White Silkie hen.

Handling your chickens

In order to maintain her place in the pecking order, a hen will disguise the fact that she is not feeling well. Handling on a regular basis is very important as it is the only way to tell if a bird has lost weight or not. Not only loss of weight, but excess weight gain, can be assessed by feeling the pin bones either side of the vent (anus). They are sharp if the bird has little fat and well padded if too fat, which will adversely affect laying.

The distance between them will indicate if the hen is laying: three vertical finger widths between the bones indicates production and less than two, non-production.

How to pick up a chicken

You can pick up a hen by spreading both hands around her body to cover her wings, then gently lift her up. Move one hand so that her breast rests on one outstretched palm, her legs between your first/second and third/fourth fingers. Take the weight on your forearm and hold her close to your body, her head pointing towards your armpit, leaving your other hand free to inspect the bird. Be firm but don't squeeze the body tightly as this may temporarily harm the breathing mechanism. This principle of holding applies to all species and all sizes of poultry - the bird is balanced and comfortable and the mucky end is away from you. The hip of any poultry will dislocate with horrifying ease if a bird is held by one leg. NEVER hold them by the legs upside down.

You can practise after dark when the hens should be on their perches. If you move quietly and slowly with a very dim torch, talking to them all the time, you will not startle them and you can then pick one hen off the perch with both hands around her wings and body, facing towards you. Then continue by sliding one hand under her as above. To catch a hen during daylight hours, I recommend a fishing landing net as this can (with practice and the aid of a fence or wall) be dropped over the hen and you can then pick her up. This is much less stressful for both you and the hen than chasing her around the pen or garden. She will be able to run and jink much faster than you!

Hens can become really tame if handled properly on a regular basis.

Opposite: Handling the correct way.

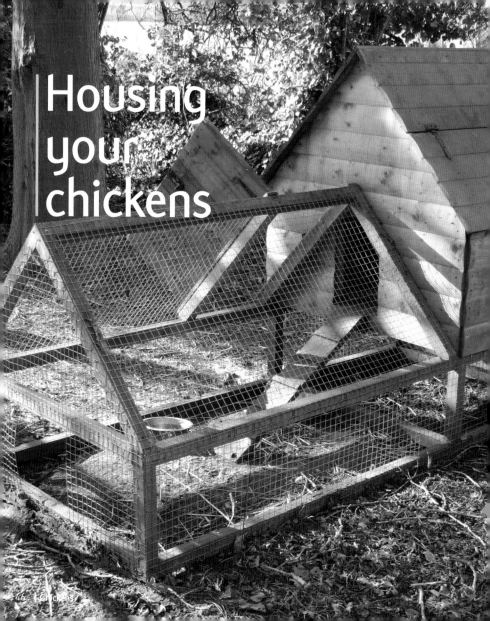

Housing your chickens

If housing is bought from a reputable manufacturer and meets all the requirements set out below, then that may be the quickest and easiest method of housing your birds. If you wish to make housing yourself, keep to the basic principles and remember not to make it too heavy, as you may want to move it.

Remember to make the access as easy as possible for you to get in to clean, catch birds, or collect eggs. Occasionally, second-hand housing becomes available, but you should beware of disease, parasites, rotten timber, and inability to transport in sectional form.

Space

Floor area should be a minimum of 30cm x 30cm (12in x 12in) per bird (large fowl) or 20cm x 20cm (8in x 8in) for bantams. If you can give them more space then so much the better, bearing in mind they will be spending time in the henhouse sheltering from the rain and wind and sun.

If a sliding or hinged roof is incorporated there is no need to have the house high enough for you to stand up in. It is useful to have a free-range house with a solid floor raised off the ground by about 20cm. This discourages rats and other vermin from hiding under the house and can make an extra shelter or dusting area for the birds. They are liable to lay under the house if their nestboxes are inadequate.

Ventilation

Correct ventilation is vital to prevent the build-up of bacteria, condensation and ammonia. Ventilation should be located near the roof to ensure there are no draughts. It can be more difficult keeping the house cool than warm.

Nestboxes

These should be located in the lowest, darkest part of the house as hens like to lay their eggs in secret places. Size for large fowl is 30cm x 30cm (12in x 12in) or 20cm x 20cm (8in x 8in) for bantams, with one nest box to four hens.

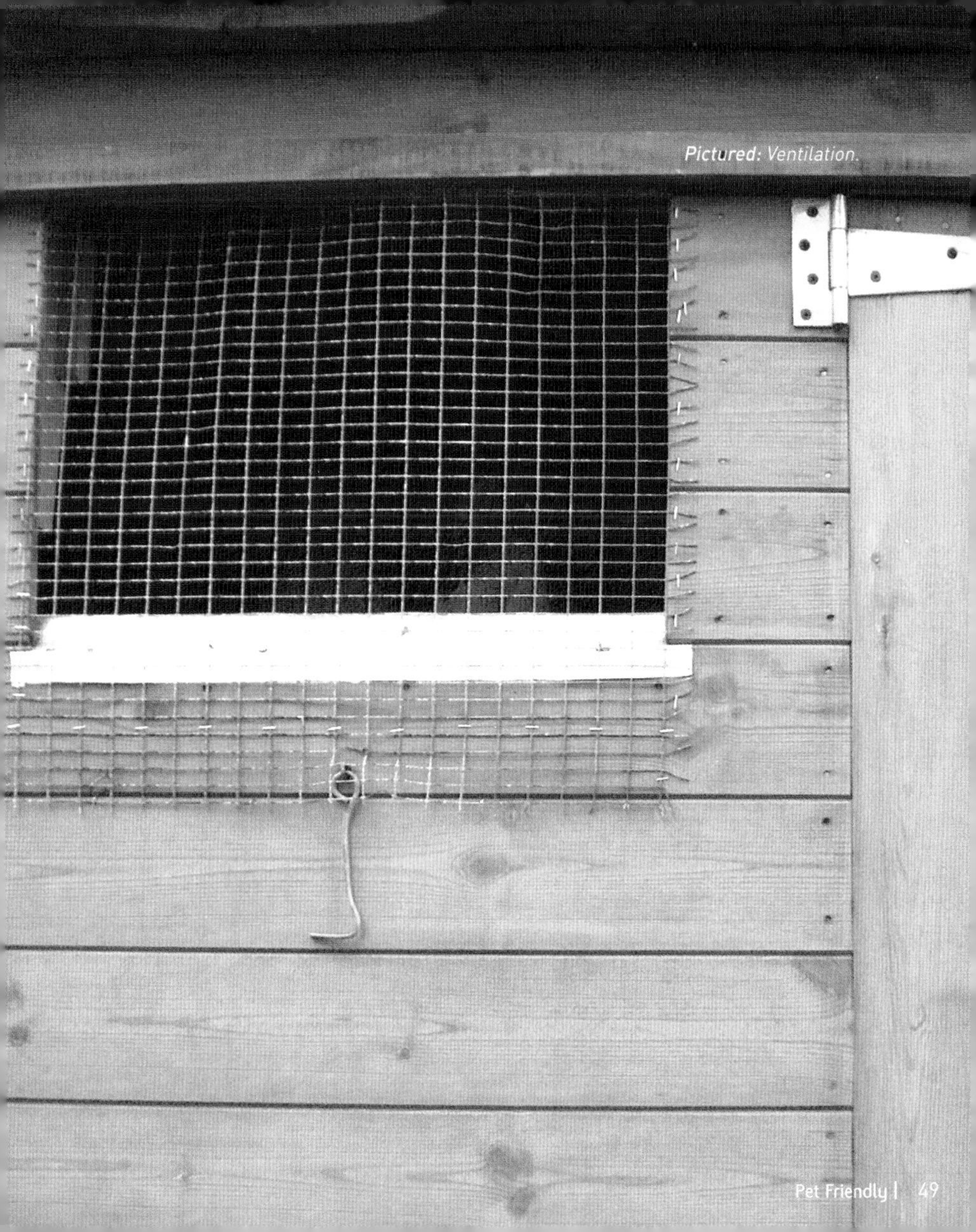

Pictured: Ventilation.

Perches

Even for bantams, perches should be broad – at least 5cm (2in) across, with the top edges rounded. They should be the correct height for the breed so they can get on them easily and have room to stand up on them, but heavy breeds should have low perches to avoid bruising on descent. Perches should allow a minimum of 22cm for large fowl and 15cm for bantams. Allow 30cm between perches. Make sure they are higher than the nestbox otherwise the hens will roost in the nestbox. Perching in birds is achieved when the bird crouches and tendons in the legs are passively tensed by the flexion of the joints, automatically clamping the digits around the perch.

Pophole

This is a low door so that the hens can go in and out of the house at will in daylight. The most practical design has a vertical sliding cover which is closed at night to prevent fox damage. The horizontal sliding popholes quickly get bunged up with muck and dirt and are difficult to close.

Security

The house must provide protection from vermin, such as foxes, rats and mice. Square 2.5cm mesh over the ventilation areas will help keep out all but the smallest of vermin. You may need to be able to padlock the house against two-legged 'foxes'– sad, but true.

Litter

Wood shavings make the cleanest and best litter for livestock. Straw may be cheaper but check that it is fresh and clean, not mouldy nor contaminated by rodents or cats. Do not use hay due to harmful mould spores which will give the hens breathing problems. Litter is used on the floor and in the nestboxes.

Cleaning

Weekly cleaning is recommended, replacing litter in all areas. There are disinfectants available which are not toxic to the birds and will destroy most bacteria, viruses and fungus harmful to poultry. If access to the hen house is easy, then cleaning is more likely to be done regularly.

Opposite: Hexagonal hen house.

Free-range
Area, vegetation

The definition of free-range is that hens have access to ground covered in vegetation (normally grass) during daylight hours. They will need a minimum of a one square metre (11 sq. ft) each. You can let the birds out when you are at home – free range does not necessarily mean total freedom.

What tends to happen is that hens are kept in one area and, within a very short space of time, eat the grass and dig up the roots, leaving nettles and docks and mud. To prevent this, 2.5cm (1in) square mesh can be laid over the grass. The hens can walk on it and eat the grass, but not dig up the roots. They will appreciate a separate dust bath if this method is used, as the netting will prevent this natural activity.

If you are gardening, the hens will love to help you find worms and insects, but they are best let out under supervision as they have a tendency to try and re-plant everything. Bantams do least damage.

Most poultry will avoid eating poisonous plants due to their bitter taste, but chickens are at risk from laburnum seeds, potato sprouts, black nightshade, henbane, most irises, privet, rhubarb leaves, rhododendron, oleander, yew, castor bean, sweet pea, rapeseed, corn cockle, clematis, common St.

Opposite:
Small flock
free-ranging.

John's wort, meadow buttercup, vetch, ragwort and some fungi. Blue-green algae is quickly fatal, so water containers should be kept clean, especially in hot weather, and access to stagnant water should be prevented.

Unless the covered run is a large area, don't attempt to plant shrubs inside it as the hens will soon dig these up. Clematis, honeysuckle, berberis, pyracantha or firs can be grown on the outside of the run both for shelter and to enhance the area.

If you want part of your garden weeded by your chickens you will need a portable combined house and run, known as a fold unit, which can be regularly moved to a fresh piece of ground.

Limited range

Types of run, aviary

Portable ark: This should be moved daily, and is good if you only have a small area of grass. By the time the run gets back to its original spot, the grass, which will have been fertilised by the hens, will have recovered.

Opposite: Small movable coop suitable for bantams.

Fold unit: This gives the birds fresh ground regularly. Some have wheels, which makes moving easy for anyone. When using a movable pen and moving it (ideally) on a daily basis, it is useful to have feeders and drinkers attached to the unit to avoid having to take the equipment out and put it all back again.

Permanent run: You can build this yourself, preferably on a grassed area. Make it as large as you can and divide it into two, so that one side can be rested on a regular basis. A large run should be netted over the top to prevent wild bird access. If this is done, then the feeder and drinker can be put outside, otherwise feeders and drinkers should be inside the house to discourage wild birds.

They are a threat in terms of bringing disease to your chickens and will also take a huge amount of the chicken food. Magpies will quickly learn to take eggs, even from inside the henhouse.

The ground in a permanent run that has been denuded of grass can have litter added to avoid mud. Pea gravel is good as it can be hosed down, wood chips (not bark as this harbours harmful mould) or rubber chips maintain free drainage.

Aviary system: This is another method of keeping poultry, but it requires more land. The principle of an aviary is that it is spacious, at least one side has mesh, the roof is solid and the walls can be either solid or mesh. The birds are protected from the elements and have plenty of fresh air without contamination from wild birds. The floor can be covered in straw, sand, gravel or wood chips.

There should be a shelter with perches at the back for roosting, plus laying boxes, feeder and drinker. Furniture, such as branches to climb on, can be provided for entertainment. Plants need their roots outside the aviary in order to survive the scratching.

Chicken-proofing your garden: fencing: Hens are very good at getting to where they think the grass is greener, by flying over, getting under, or squeezing through a very small hole in a fence or hedge. A depression in the ground where you have put a fence is a typical place for escape – and they can be very persistent.

Fencing needs checking on a regular basis for escape routes. Cut-up wire coathangers can be bent and used as pegs to hold netting to the ground.

Wing clipping

You can stop chickens flying over a fence by clipping the feathers of just one wing in order to unbalance the flight process. Heavy breeds may never need this, light breeds are more flighty and may need it done annually after the moult.

Wing clipping is done by using the small covert feathers as a guide to ensure that the cut is away from the skin and enough of the flight feathers are removed, cutting across the primaries with scissors. It is similar to us cutting our fingernails. The feathers will not grow back until the next moult which is usually in autumn. It is important that before clipping, the primaries are fully grown through and the blood in the quill has retreated with the white sheath fallen off which avoids the feather stub bleeding when the primaries are cut. This is easy to see by looking at the underside of the primary feathers near the skin.

If you are in any doubt about being able to perform this correctly, consult your veterinary surgeon or an experienced chicken keeper.

Types of vermin

Mice: Attracted by food. Trap or prevent access.

Rats: Attracted by food and eggs. Poison or trap.

Grey squirrels: Attracted by food and eggs.
Poison or trap.

Weasels and stoats: Attracted by the birds. Trap or
prevent access.

Mink: Attracted by the birds. Trap or prevent access.

Foxes: Attracted by the birds. Prevent access.

Feral cats: Attracted by the birds. Prevent access.

Magpies, crows, jackdaws and rooks:
Attracted by food and eggs.
Prevent access.

Animals which will take hens or eggs but are protected by law in the UK include owls, buzzards, hedgehogs, and badgers.

Predators in the USA also include snakes, racoons, coyotes and birds of prey. Predators in Australia include snakes, carnivorous marsupials, carnivorous lizards and birds of prey. Local laws will need to be checked for information on protection of certain species.

You can protect your hens from rats and mice by setting traps. A multi-catch mouse trap can be used in the hen-house; rat traps should be set in tunnels. If you decide to use poison, make sure it is not accessible by pets or children.

Foxes, cats, mink and squirrels can be kept out by the use of fencing or covered runs. Fencing needs to be over 2m high or with an overhang, plus netting on the ground to prevent digging, or with an electric fence single wire at the base. Commercial electric netting is fine for keeping the chickens in but will not keep a fox out as he can jump over it.

Magpies may try to gain access through the pophole, so pin vertical strips of black binbag over it. The hens will push their way in but the magpies will not like the movement of the strips.

Left: Brown Rat.
Right: Jackdaw.

Feeding your chickens

In order to produce good-quality, healthy eggs, chickens need to have a regular supply of nutritious food as well as vitamins and minerals. This is best supplied through feeding proprietary chicken layer pellets, together with a little grain such as whole wheat.

You should ensure your chickens are fed the right food for their age, so use chick crumbs to six weeks, then growers pellets to 18 weeks, then layers pellets, for example. Cheap feed will be of poorer quality.

It is illegal in the UK to feed chickens scraps which have come from your kitchen. This is to prevent disease. You can feed your hens surplus greens directly from your vegetable garden, though.

Opposite:
Top: Whole wheat.
Bottom: Layer pellets.

The best commercial feed is pellets. Feed is available as meal/mash but this may be wasteful and sticks to the beak, quickly making any water foul.

If birds have access to grass they will not need extra greens but if, in the winter, there is not enough grass, hang up cabbage stalks, nettles or brussels sprout plants in their hut.

Clean water and mixed grit should be available at all times. Hens dehydrate quickly, while flint (or insoluble) grit is needed to assist the gizzard in grinding up the food, especially hard grain. From four weeks before laying commences, oyster shell or limestone grit should be freely provided to help with the formation of egg shells.

Light breeds start to lay at about five months and heavier breeds at six months plus. Large fowl will eat 110-170g per day, bantams need around 50-85g, according to size. Wheat (and a little maize in cold weather only – it is too heating in warm weather and can lead to feather-pecking) can be offered as a scratch feed to keep the birds active.

Keep feed in a vermin-proof and weather-proof bin to keep it fresh. Check the date on the bag label at purchase as feed will last only three months before the vitamin content degrades to an unacceptable level.

Equipment: feeder, drinker

Commercially-produced specialised feeders and drinkers are readily available, and help keep feed and water clean by not allowing the birds to defecate in it.

The feeders allow free feeding, which means that more feed is put in than the chickens can eat in one day. Hens are good at eating what they need for egg production and only get too fat if too many grains are fed.

The feeder can be hung up at hen shoulder height and is best with a lid on to prevent perching and therefore mucking in the feed. If rats are a nuisance, the feeder should be removed at night. Some metal feeders respond to the weight of the bird. She stands on a treadle which opens the feed area, making it rat-proof!

Drinkers can be of varying sizes and made of plastic (cheaper but not very long-lasting) or galvanised metal (more expensive but last for many years). The larger sizes can be very heavy when full and difficult to keep ice-free. In severe weather, a simple plastic washing-up bowl with slightly sloping sides will allow ice to be tipped out and may need to be refreshed several times a day.

Pictured: A type of drinker.

Caring for your chickens

Caring for your chickens

Daily tasks

Give the hens as much time as you can as you will enjoy your hobby more. But it can take just a few minutes daily to do the important things.

- Let birds out at dawn in winter and about 7am in summer.

- Change the water in the drinker.

- Put food in the trough or check the feeder has enough food in it for the day.

- Collect eggs.

- Add fresh litter if necessary.

- Scatter a little whole wheat either in the house or in the run.

- Observe birds for changes in behaviour which may indicate ill health.

- Shut pophole before dusk, checking all birds are in the house.

Weekly tasks

Clean nestbox and/or droppings board or floor. Putting cardboard under the shavings makes this easier, and it will compost down.

- Scrape perches.

- Wash and disinfect drinker and feeder.

- Check mixed grit is available.

- Handle every bird to check weight and condition.

- Check the hut after dark for red mite.

- Deal with any muddy patches in the run.

- Check beak and claws and, if necessary, trim to the correct shape with a file.

- To trim beak and claws, one person holds the bird (see Handling, page 44) and the other uses a nail file to smooth the beak and claws to a natural shape. This may need doing once or twice a year.

If you provide a droppings board under the perches which can be removed easily it will help keep the floor of the house cleaner, as hens produce two thirds of their droppings at night. You can also check the droppings for colour and consistency (as a guide to health) more easily.

Left: Circles= normal ordinary droppings, central are the caecal droppings, about 1 in 10 ratio to normal ones.

Right: Australorp hen.

Going on holiday

A day out can be easily covered by checking the birds before you leave and, if you will not be back before dark, getting a friend to close the pophole or fit an automatic closer which works on light levels.

It is not acceptable to leave hens for a weekend without them being checked, even if food and water is provided, so provision must be made for someone to do this.

If you go on holiday, someone will need to look at your birds at least once a day, feed and water them, collect the eggs and open and shut the pophole.

Your neighbours may be the answer, but in some areas you can board your hens as you would a dog or cat. There are also house and pet-sitting services which may be able to help.

Opposite:
Hens behind
fencing.

Understanding
chickens

Understanding chickens

Chickens have a poor sense of taste and smell, but they are sensitive to texture. They use shadows to spot potential food items on the ground and anything falling or moving is immediately investigated. All birds have colour vision and hens are particularly attracted to red, hence the red bases of chick drinkers. Unfortunately this also means that fresh blood is attractive, which can lead to cannibalism, no matter how large the free-range.

Hens are not good gardeners – they seem convinced that everything you have planted is upside down! What they don't scratch up, they dust bathe in, to help remove external parasites. They will scratch dust (or dry ashes or sand) and flip it all over themselves with their wings, then stand up and shake it all out.

Hens will also sunbathe, lying stretched out with leg and wing extended, looking as though they are dead! They do this because the sunlight changes the oil from their preen gland (just above the tail and the size of a biro-point), which they have wiped over their feathers when preening, into vitamin D, which is important for bone strength.

Anything overhead is a potential predator. One hen will often alert the flock to a sparrowhawk or buzzard and then there is a general warning buzz.

Hens do not see well in the dark, so they feel safe in a dark enclosed space at night, putting themselves to bed before twilight. They do not herd well, even in daylight, and two long bamboo canes can work wonders in getting them all to go in the desired direction.

Good feather condition is important for insulation, and moulting occurs once a year, generally in the autumn, taking two to three weeks.

Correct nutrition is really important while this is going on. The colour of feathers can fade in sunlight over time.

Feathers are arranged in the skin in tracts (lines) with areas of no feathers in between. In order to maintain body temperature, the feathers may be raised or lowered by muscles under the skin to trap air between the feathers and the skin for insulation and for indicating mood and behaviour.

The shape of the chicken does not change if it has gained or lost weight, so regular handling is important to check your chickens' health.

Chickens have strong survival instincts. They need routine for stability and anything new needs either avoiding or investigating. They all have individual characters.

If chickens are free-range and the hen house is to be moved, do it over small distances so they follow it, to avoid them roosting on the ground where it was before.

In harsh weather and snow, chickens will not immediately go outside as the ground has changed colour and is therefore not to be trusted. If some feed is scattered on the snow they will soon get the idea.

Talk to your chickens. They will recognise your voice and your clothes. But remember they have colour vision, so sudden changes of colour of clothes can startle them.

Chickens can be taught tricks with the aid of food. For example, they will learn to peck at different coloured containers to find hidden food.

Chickens wish to hide their eggs, which is why the nestbox is placed in the darkest part of the hen house. Some hens will go off and lay their eggs in the most secret place in the garden, so keep watch and listen for hens returning (and the triumphant clucks) if the number of eggs seems to have reduced.

They are also good at jumping and will work out how to use the roof of the hen house to launch over a fence, despite a clipped wing, so bear this in mind if you have a persistent escapologist.

Hens can be immensely stupid in rain. The older ones will quickly decide if it is only a shower or a lasting downpour and run for home, but young birds are liable to stand around in the wet looking miserable.

They will become cold, especially those with thin skulls and crests, such as Polands, and those with woolly feathering, such as Silkies.

Introducing new birds

The word "henpecked" is so much part of our language that most would use it without thought to its origin, but the pecking order, as we have seen, is a vital component in maintaining the stability of the flock in hens.

With or without a cockerel, hens have a strict hierarchy and territory which only changes if a bird is removed or is sick.

How then do you return a recovered bird to a small flock? With great care and supervision, as a chicken's memory is very short and the whole order has to be reshuffled. The recovered bird must be fully fit in order to reclaim her place. Hens may fight to decide the natural order if there is no cockerel.

Should you wish to add three or four new laying hens to an established flock, the least traumatic way to do it is to put old and new into a fresh henhouse so none has established a territory.

If that is not possible, let the residents get used to the idea of newcomers by partitioning the hut and run (or put a smaller hut and run within the compound) for a couple of weeks so that the newcomers can be seen but not attacked.

How an egg is produced

Laying is hormonal and influenced by light levels. An egg is created by the production of yolk in the liver which then travels in the blood to the ovary. Chickens have only the left ovary functional.

Ovulation occurs when one ovum matures and drops into the top of the oviduct, a funnel-shaped tissue called the infundibulum. The egg then passes through the different areas of the oviduct (the addition of the shell taking the longest time) over a total period of 25 hours, so there can never be two eggs from one hen in one day.

The hen will ovulate 30 minutes after laying, but eventually it will be dark when this should happen, in which case she misses producing an egg the next day. This gap marks the end of one clutch. Hybrid hens have been selected to produce an egg in a slightly shorter time, hence they lay on consecutive days for longer.

Brown eggs have pigment placed on the outside of the shell which can scratch or wash off. The only egg which has pigment all the way through the shell is from the Araucana, and is blue/green.

There are three types of white in a fresh egg – thin white, thick white and the chalazae, which are strings of very thick white which keep the germinal disc on top if the egg is rotated. You are unlikely to be able to differentiate the types of white in an egg from a supermarket as these are not fresh (like a new-laid egg is fresh) and can be several weeks old.

Eggs should be kept in the refrigerator if ambient temperature is above 10ºC (50ºF) to help prevent bacterial growth and will keep for several weeks like this.

Once people start keeping hens, many want to breed chicks. But remember that the sexes will be about 50% hens and 50% cockerels, so you will need a plan to get rid of the inevitable spare males.

Broody hens

A broody hen has the instinct to sit upon eggs, keeping them warm and incubating them until they hatch. Hybrids have been selected so that broodiness is reduced but they can still sometimes go broody. You will find that one hen will insist on staying most of the time in the nestbox and may well try and peck you when you are collecting eggs.

To check if she is really broody, gently slide your hand under her, palm up, and if she "cuddles" your hand with her wings, then she is serious.

The best way to reset her cycle is to construct a "sin bin". This could be as simple as a rabbit hutch, but should really have a small mesh floor so that she does not want to think of a nest, plus has food and water available. It will take about a fortnight for this to be effective, harden your heart and do not let her out before this time as she will go straight back to the nestbox and begin sitting again. After two weeks in the sin bin, integrate her carefully back into the flock (or place the sin bin in the henhouse if there is space) and she should then begin to lay properly again after that quite quickly, therefore earning her keep. The next time she looks as though she is going broody put her in the sin bin straight away for a few days. This may be enough as the hormones are not at full power at the beginning of the broody cycle.

Selling surplus eggs

In the UK, you are allowed to sell surplus eggs directly to the end user. Regulations on selling eggs begin if they are sold to a third party or there are over 350 hens kept.

Breeding chickens

In order to hatch, eggs need to be fertile. A cockerel mates with the hens during the Spring as the days get longer. If it is not possible for you to keep a cockerel (the noise they make is irritating to some people), you may be able to purchase fertile eggs and either put these under your broody hen or artificially incubate them.

A strong cardboard box makes an ideal nest. A broody stays broody until she hears chicks cheeping, so if she is sitting on hen eggs, this takes 21 days.

Natural hatching

Natural hatching under a broody is the ideal way to raise a few chicks. It is, however, essentially dependent on having a broody or broodies at the same time as the eggs you want to set. Silkie crosses make the best broodies (either x Wyandotte or x Sussex) and most bantams will also go broody. A small pen of these can be bred alongside the purebreds or layers.

Use a disposable cardboard carton (for hygiene reasons) for the nest, making sure it is in a fox-proof area, in a quiet spot away from other stock. Use shavings for litter (hay produces harmful moulds), dusted with a pyrethrum-based insecticide, with good ventilation near the top.

Broody birds of any species which sit beside each other spend their whole time stealing each other's eggs and generally ruining a hatch. The broody hen is best moved gently in the dark to her new nest to keep her sitting and left for a day or so on just a few unimportant (or pot) eggs to ensure she is still serious and not upset by the move.

Some lighter breeds are best left where they decide to sit as if they are moved they will go off being broody. When a hen is broody, a bare patch of skin on her breast (brood patch) is in contact with the eggs to maintain temperature and humidity.

Once the broody is steady, put the eggs you want to hatch under her very gently, preferably at night, and remove the others. Set an odd number of eggs as these fit better into a circle. If you want to set more than one broody at a time, make sure that you set the eggs the same day so that they hatch all together, or keep the broodies out of sight and sound of each other. The noise of cheeping will make the other broody get off her eggs and help with the hatching ones.

The broody hen should be taken off the nest once each day to feed, drink and defecate at about the same time.

In the last two days before the eggs are due to hatch, the hen should not be disturbed. A chick drinker (small so they do not drown) and chick crumbs should be left within her reach. The hatch may take two to three days to complete, but the early chicks need to be able to feed. Any eggs left after three days should be gently shaken beside your ear: if they rattle, be careful they do not explode!

A rattling egg is a rotten egg and the smell will be difficult to get rid of.

The hen needs to bond with her chicks and turn her sitting instinct into the more aggressive and protective maternal instinct, but gently removing empty shells in the dark will mean the other eggs can hatch with more space.

If you wish to amalgamate broods under one broody, you have just 24 hours as a window. After that the hen knows which chicks are her responsibility and will attack any "intruders". This comes with the territory, so knowing the behaviour will help prevent disaster. There are some broodies who will gladly welcome any chicks, but they are unusual.

If you buy in day-old chicks for a broody, give them a drink by dipping their beaks in tepid water, then place them in a smallish cardboard box so they stay warm and put them within sound of the hen for about an hour. Then conceal them in your palm and place them gently under her, removing the eggs at the same time. Best done in the dark, but it will depend on what time of day you get the chicks. Remember you have only 24 hours if you wish to add more: she cannot count but she has colour vision and can tell the difference between chicks.

Artificial hatching

Small incubators are regularly used by many poultry keepers, the advantage being that incubation conditions are instantly available at the flick of a switch.

Best results will be obtained with eggs which are between 24 hours and seven days old and which have been stored in a cool (10°C) place and turned daily. Mark the eggs in pencil with the date and breed and keep records. Any dirt on the eggs can be scraped off with a dry potscraper, the ideal being to have clean eggs in the first place. If eggs do have to be washed, use water warmer than they are to ensure that the membrane under the shell expands, keeping bacteria out (cold water makes it shrink, drawing bacteria in) plus a poultry disinfectant such as Virkon or F10.

The same disinfectant can be used with safety to clean out incubators after a hatch. This is most important for the success of future hatches as the bacteria and debris produced by a hatch is phenomenal.

Run an incubator for a few days, checking the temperature with another thermometer. Do not add any water – it seems to be a common misconception that water needs adding during the incubation process. Try and site the incubator in a place which does not vary much in average temperature, such as a spare bedroom, but avoid direct sunlight. During the incubation process the eggs must be turned in order for the embryo to develop normally (the broody does this by instinct). If turning by hand, do so at least twice a day and turn the eggs end-over-end so that the chalazae (strings which hold the yolk stable) do not wind up, potentially damaging the embryo, or mark the eggs and turn them from one side then back to the other, not continuously in the same direction.

If the incubator is an automatic turning one, turn off the mechanism two days before the hatch date, or stop turning them by hand at that time.

The best system is to have a separate hatcher so that eggs can be set weekly and do not actually hatch in the incubator, keeping it cleaner. Eggs should be moved to the hatcher two days before they are due to hatch.

Hot water can be added in the base of the hatcher when the eggs start to pip (the diamond-shaped start of the shell breaking) to increase humidity, keeping the shell membrane moist. The chick pecks its way out of the broad end of the egg by means of the egg tooth (this falls off soon after hatching) which is on the end of its top beak.

Chicks may take two days to hatch or they may all hatch at once. The latter is better, but not always possible. Most small incubators have a window in so that you do not have to take off the top to see inside. If you do not have a separate hatcher, it is better to fill (or part fill) an incubator, hatch the eggs, clean it out and start again, avoiding the build-up of harmful bacteria.

In order to make best use of incubator space (and broody hens for that matter) the eggs can be candled after seven days' incubation.

This involves holding a bright torch to the broad end of each egg in a darkened room. Obviously white-shelled eggs will be easier to see into than dark brown eggs.

If the egg is infertile you will be able to see just the shadow of the yolk. Rotate the egg slightly to make this move within it. If fertile, a spider-shape of blood vessels will be seen with the heart beating in the middle. If there is a ring of blood vessels with none in the centre, the germ has died for some reason. The air sac gradually gets larger as hatching date approaches and sometimes the chick can be seen bobbing away from the candling light.

Take the chicks out when they are dry and, keeping them warm and out of the light, transfer them to their rearing quarters and dip their beaks in tepid water.

Incubators are available from poultry suppliers and the capacity ranges from three eggs up to several hundred. Cost begins at around £50 ($80).

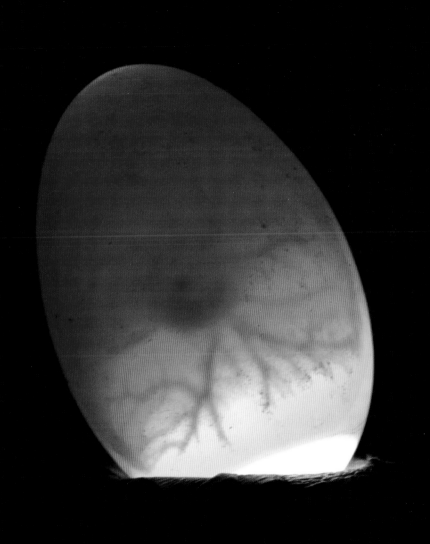

Rearing your chicks

Rearing your chicks

If you have a broody hen to rear for you, all you will need to provide are chick crumbs, water and shelter against wind, rain and sun, preferably with a wired-over run so that magpies and crows cannot take the chicks or steal the food.

Chick crumbs need to be in a container which the hen can neither tip over nor scratch them out of. The water container should be such that the chicks cannot drown. Put some grain feed for the hen out of reach of the chicks. She may break the grain into small pieces for them.

Leave the hen with the chicks for about four weeks and then take her away, rather than the chicks. They can be transferred to a larger house and/or run when they are about eight weeks old.

Incubated chicks need a heat lamp to keep them warm, preferably one with an infra-red ceramic bulb so that they have heat and not light. The size of the bulb will depend on the number of chicks, with a 100 watt one being sufficient for a few chicks and a 250 watt one needed for 50 chicks. The heat without light avoids feather pecking as they then have natural light and darkness to maximise body and feather growth.

On a small scale, use a large rectangular cardboard box and change this for each hatch. It needs to be rectangular so that the lamp is at one end and the chicks can regulate their own temperature by moving away from the lamp.

Turn the heat lamp on two days before the chicks are due to hatch. It should be far enough off the shavings so that the temperature under it is about 39°C. If the chicks are too hot they will scatter to the edges, panting. If they are too cold they will huddle in the middle, cheeping loudly. The ideal is a small empty circle just under the lamp.

Transfer the chicks from the incubator when they have dried and fluffed up. Dip their beaks in tepid water and place them under the lamp. If they are thirsty and you give them cold water, the shock can kill them.

Provide chick crumbs in a scratch-proof container. Chickens have the instinct to scratch the floor with their feet from one day old and if their food is in a bowl, they will scratch it all over the floor and waste it. They can stay in this area either until they outgrow it or they are weaned off the heat lamp, at about six weeks. The lamp can be gradually raised and then turned off in the middle of the day if it is hot outside, but it must be put back on at night.

The chicks should be well feathered by this stage and able to keep themselves warm, the lighter breeds feathering up quicker than the heavier ones.

If you have used a heat lamp that also produces light, try leaving this off at night when the birds are about four weeks old if it is reasonably warm outside. That way they get used to the darkness, otherwise they could panic and smother when you eventually put them in a dark hut.

Health care

Chickens are prey species and so hide their symptoms if they are sick. A change in behaviour is often the first sign that the observant owner sees.

Signs of poor health

A sick chicken will stand with its feathers ruffled and its eyes closed to conserve energy. Veterinary attention is needed if you see the following signs:

- Foam in the corner of the eyes, swollen sinuses and nasal discharge: likely cause mycoplasma.

- Difficulty in breathing: likely cause respiratory disease.

- Lethargy, standing around with eyes closed: likely cause internal problem.

- A change or blood in the faeces (remember the caecal faeces are different): possible cause coccidiosis.

- The comb goes pale: likely cause red or northern fowl mites.

- The legs become rough with white raised areas: likely cause scaly leg mite.

- Lameness.

- Unusually shaped eggs.

Parasites

Internal parasites: These are common in hens which are outside during the day. They are always on the look-out for insects and worms and some of these can contain harmful parasites. It is easy to control these by giving a the only licensed poultry wormer called Flubenvet to the hens in their feed. If the birds are in an ark and moved regularly, they should be wormed 2-3 times a year. If they are on the same ground all the time, this interval should be shorter as they may be re-infecting themselves frequently. Flubenvet has nil withdrawal time for eggs, so it can be used at any time. It is obtainable from your vet or agricultural merchant. Some of the internal parasites can't be seen with the naked eye, so worm your birds on a regular basis.

External parasites: There are several types, all of them need dealing with as dustbathing will only remove a few:

- Lice live on the bird, are yellow, about 3mm long and lay eggs (nits) at the base of feathers, usually under the tail, so this is a place to look on a regular basis. Control is with louse powder based on pyrethrum. Not life-threatening, but reduces production.

- Mites: a) the red mite is 1mm long, nocturnal and sucks the blood of hens at night, living in the hut during the day. Hens become anaemic with the comb and face pale and can die. Red mite can be controlled by spraying the hut with various licensed products when the birds are outside, but the nooks and crannies, especially under a felt roof, are difficult to get at. Careful application of a blowtorch is just as effective, although time-consuming. Either treatment will probably need several applications. The red mite is grey when it is hungry and will take a meal off a human if it gets the chance. They can live for up to a year without feeding – beware secondhand henhouses!

- If the hens refuse suddenly to go into the hut at night, suspect red mite.

- b) the northern fowl mite is a relative of the red mite but lives and breeds on the bird all the time. On a white bird, this is easy to see as a dirty mark on the base of the feathers or under the tail. Around the vent is the most common place to find these so, again, check here regularly. The birds are pulled down within a few days of being infested due to anaemia and can die. The life cycle in warm weather of both types of mites can be as short as 10 days, so vigilance is really important. Treatment is pyrethrum based louse powder or an avermectin (ask your vet) as nothing is licensed for northern fowl mite.

- c) scaly leg mite: this burrows under the scales of the legs making raised encrustations and is very irritating for the bird. Treatment is dunking the affected legs in surgical spirit once a week for 3 weeks but scales moult like feathers so will not look normal until the next moult.

Tips on preventing disease

- Poultry healthy respiratory system depends so very much on good ventilation that disease can be prevented: damp stale air will quickly cause problems, so good ventilation at the top of the hen house with no draughts is best.

- High levels of ammonia from the litter stop the removal of phlegm and so invite bacteria and viruses to multiply, so regularly cleaning out is important.

- Hens are omnivorous and enjoy catching and eating mice, but the disease risk is high from rodents.

Old long grass needs avoiding as this can impact and kill, as can polystyrene (e.g. ceiling tiles) which they just adore to peck at, or pieces of plastic string. Keep the hens' area free of litter.

- Hens have evolved to scratch around in the dirt, but over a wide area. Muddy areas encourage harmful parasites to breed so put down slats or move the hut more regularly to avoid mud.

- Feathers are good insulators and it is sometimes harder to keep birds cool in summer than warm in winter. Birds that are too hot will hold their wings out from their body and pant, they can be sprayed with water to cool down and shade needs to be provided.

- Hybrids are automatically vaccinated against Infectious Bronchitis virus. Other backyard hens are only vaccinated if a disease has been diagnosed by your vet. There is a small risk mixing vaccinated and unvaccinated birds, but chicken vaccines are only effective in the very young, so vaccines do not work in adult chickens.

- Use cider vinegar (10ml to 500ml water, plastic drinkers only) for one week a month in order to help keep most pathogens at bay.

Biosecurity for free-range poultry

- Isolate new stock for two to three weeks.

- Isolate birds for seven days after taking them to an exhibition.

- Change clothes and wash boots before and after visiting other breeders.

- Change clothes and wash boots before and after attending a sale.

- Keep fresh disinfectant at the entrance to poultry areas for dipping footwear.

- Care for young birds before older birds to avoid the spread of disease.

- Disinfect crates before and after use, especially if lent to others. However, it is preferable not to share equipment.

- Disinfect vehicles which have been on poultry premises, but avoid taking vehicles onto other premises.Wash hands before and after handling poultry.

- Comply with any import/export regulations/ guidelines.

These are commonsense measures which can easily be incorporated into the daily routine.

Hygiene

- High ammonia levels in a hen house will encourage respiratory diseases.

- Poultry equipment (e.g. feeders and drinkers) should be kept clean and free of algae.

- Hand washing after handling chickens or their equipment is essential.

Chickens & the law

Local regulations and authorities in your country or state will have details on whether there is a by-law for your property or area which prohibits the keeping of poultry.

Neighbours may be happy with the level of noise from cockerels but it only takes a complaint from one person...Cockerels crow as a territorial activity, so if there is only one cockerel in hearing distance, there is less noise.

The following measures can be taken to reduce the nuisance of cockerels crowing:

If you are not going to breed from your birds you do not need to keep a cockerel.

During the breeding season the number of cockerels can get out of hand. Be realistic and only keep the cockerels you require as replacement stock.

Think carefully about the positioning of the poultry houses. Do not place them near to neighbours if at all possible.

Provide the birds with a house where the light entering it has been eliminated as far as possible. Always remember that the birds will require ventilation in their housing.

If possible try to explain your hobby to your neighbours and invite them round to see the birds. A gift of a dozen eggs always goes down well!

Lastly, keep your cool, listen to what is being said and try to co-operate with the local authority.

Chicken facts

Normal body temperature:
40-42°C. Birds have a high metabolic rate due to the demands of flying.

Heart rate:
120-160 beats per minute.

Normal breathing rate:
20-30 per minute.

Life expectancy:
Ex-battery hens: weeks to months

Hybrids: 3 to 4 years.

Pure breeds large fowl: 6 to 10 years.

Bantams: 8 to 12 years.

Sources of further information

Diseases of Free-Range Poultry, 3e.Victoria Roberts BVSc MRCVS, Whittet Books, 2009

British Poultry Standards, 6e. Victoria Roberts BVSc MRCVS, Blackwells, 6th edition, 2008

Poultry for Anyone, Victoria Roberts BVSc MRCVS, Whittet Books, 1998

Teach Yourself Raising Happy Hens, Victoria Roberts BVSc MRCVS, Hodder, 2009

Poultry at Home (DVD), Victoria Roberts, Old Pond Press, 1993.

Websites

www.vicvet.com: further husbandry and disease information from the author

www.animaloracle.com: decision tree for symptoms and timings of veterinary attention

www.poultryclub.org: husbandry, pure breeds, showing

www.vetark.co.uk: useful products

www.birdcareco.com: useful products

www.meadowsanimalhealthcare.co.uk (F10 disinfectant)